AF887730

THIS LITERARY ADVENTURE BELONGS TO:

www.myliteraryadventure.com.au

For Riley, April & Chelsea
&
Ted Lovegrove,
teacher at Scotch College, Adelaide 1964

First published 2017 by My Literary Adventure Inc.

My Literary Adventure Inc.
PO Box 236, Goodwood, South Australia, Australia, 5034
Phone: 61 8 8193 0450
Email: info@myliteraryadventure.com.au
Web: www.myliteraryadventure.com.au

National Library of Australia Cataloguing-in-Publication entry:

Creator:	Smith, Graham, author.
Title:	Science with The SIDEWAYS Man / Graham Smith.
ISBN:	9780994228956 (paperback)
Target Audience:	For primary school age.
Subjects:	Physics - - Juvenile literature.
	Science teachers - - Juvenile literature.
	Physics - - Experiments - - Juvenile literature.

Book Design and Graphics: My Literary Adventure Inc.

Science with 'The SIDEWAYS Man'

by Graham Smith

My favourite teacher at our school teaches us science and all our class think he is really smart. I love to learn about ***physics*** especially when his little dog, Betsy, helps him out.

I call my Physics teacher the **Sideways Man** because whenever I picture him in my mind, I only see him ***SIDEWAYS!***

FRICTION

On Monday he taught us about ***friction*** - how hard it is to drag a heavy box across the floor!

FRICTION

He showed us that with wheels underneath the box there is much less ***friction*** - as he silently rolled right out the door!

FRICTION

He taught us that ***friction*** also causes ***heat*** when movement turns ***energy*** into ***heat***.

Try it out: place a long thin stick against a soft piece of wood (or bark) and using your hands, rotate the stick backwards and forwards quickly.

The point of the stick rubbing against the wood (or bark) will cause the wood to heat and catch fire - the early Australian Aborigines were experts at it - now that is ***friction!***

LIGHT

On Tuesday the **Sideways Man** explained that ***light*** from the sun, a light bulb, or even a torch, is a form of ***energy***.
It consists of many waves of light energy, or ***beams*** travelling in straight lines that we can't see.

LIGHT

He said that when something blocks the ***light***, ***shadows*** are formed because the light waves can't bend around things. Light waves can pass through air, water, or space but they can't travel through solid objects, like your body.

LIGHT

Light rays can also be ***reflected*** by a shiny mirror which bounces light waves back to you evenly.

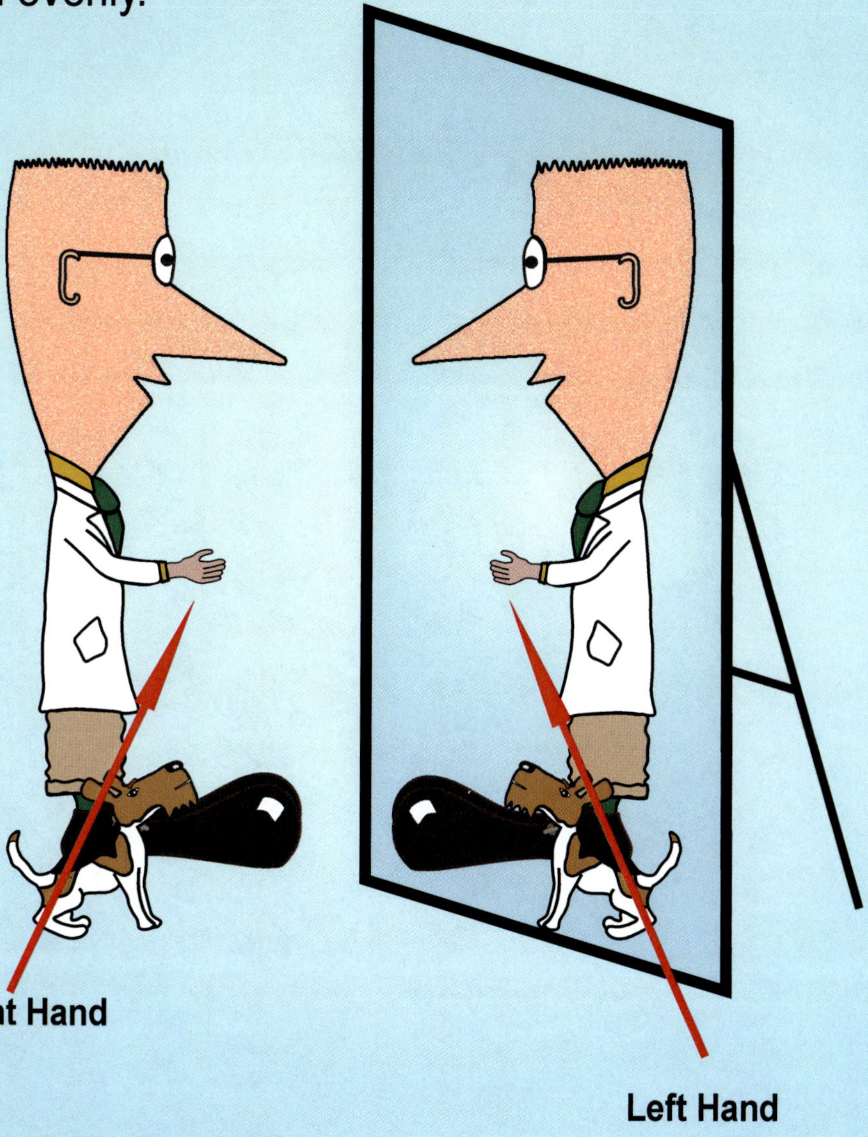

But remember, the reflection you see in the mirror is in ***reverse***.

LIGHT

Light is made up of all the colours of the rainbow.

RED, ORANGE, YELLOW, GREEN, BLUE, INDIGO, VIOLET

These colours can be seen as a ***spectrum*** when light shines through a ***prism*** or sometimes you see it through window glass, or even raindrops, to make a rainbow.

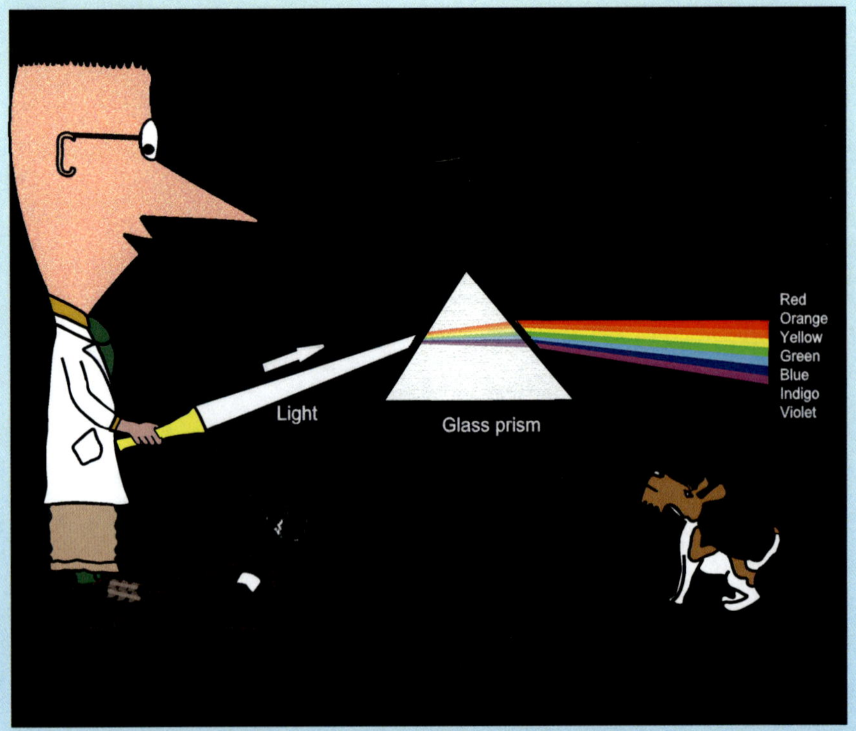

LIGHT

When ***light*** strikes an object, some spectrum ***colours*** are absorbed and some bounce off.

That is why a yellow crash helmet looks yellow and a red car looks red!

SOUND

The ***Sideways Man*** also taught us about ***sound***. Sound is made up of ***Sound Waves*** made by movements in the air called vibrations.

Our eardrums pick up these vibrations and our brains interpret the sound.

Sound Waves can be ***loud, soft, high*** or ***low***. Sounds that bounce off or reflect from hard surfaces, like a brick wall, are called ***echoes***.

The **WEATHER**

On Wednesday he taught us about the ***weather*** and that our ***weather*** (Climate) follows a cycle. He said that ***water*** is a ***liquid*** that can turn into a ***gas***, or vapour, when it gets hot enough.

1. The sun heats up surface WATER and it evaporates into water vapour, a gas. **(evaporation)**
2. The water vapour rises up into the sky where it cools and turns into WATER droplets in the form of clouds. **(condensation)**
3. When the WATER droplets become heavy enough they fall back to earth from clouds as rain. **(precipitation)**
4. The WATER is collected as puddles, lakes, rivers and the sea. **(collection)**

The **WEATHER**

He taught us how to use a home made ***Barometer*** to measure the (Atmospheric) ***Air Pressure.***

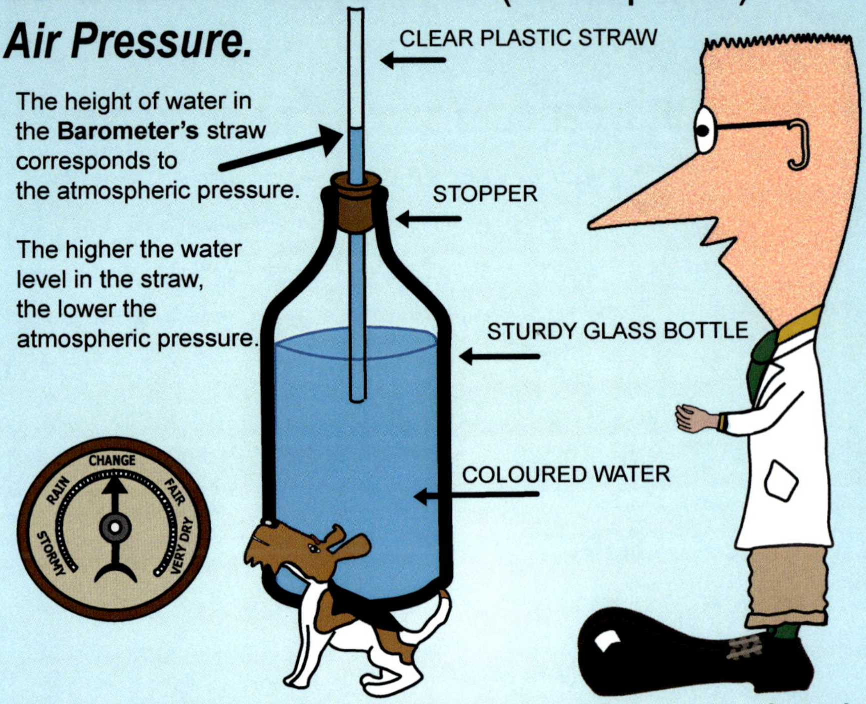

When there is ***low air pressure*** the water level in the straw rises because the air inside the bottle now has relatively more air pressure than the air outside, so water is pushed upwards. When there is ***high air pressure*** the water level in the straw drops because the air inside the bottle now has relatively less pressure, so water in the straw is pushed downwards. Generally, high air pressure means ***fair*** weather while low air pressure means ***bad*** weather.

The **WEATHER**

He also showed us how to make our own ***Rain Gauge*** to measure rainfall.

- Cut a plastic bottle in half and tape the top half inside the bottle.
- Place some stones in the bottom for weight.
- Mark a scale on the outside of the bottle.
- Fill with water to the bottom of your scale.

Put your rain gauge outside where it can collect ***rain water*** when it starts raining.

After a rain shower has finished, measure how far up the scale the ***water level*** has risen.

The **WEATHER**

He said that if you want to find out which direction the ***wind*** is coming from, make your own simple ***Wind Vane*** using streamers.

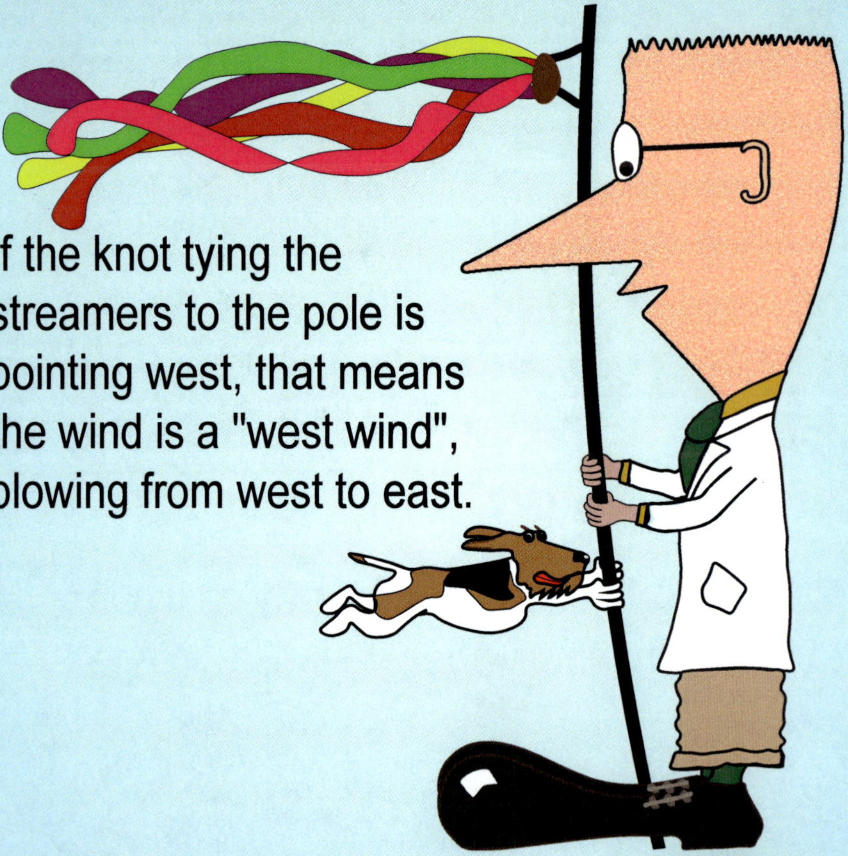

If the knot tying the streamers to the pole is pointing west, that means the wind is a "west wind", blowing from west to east.

Use a compass to see which ***direction*** the wind is coming from. The streamers will point in the direction the wind is blowing from.

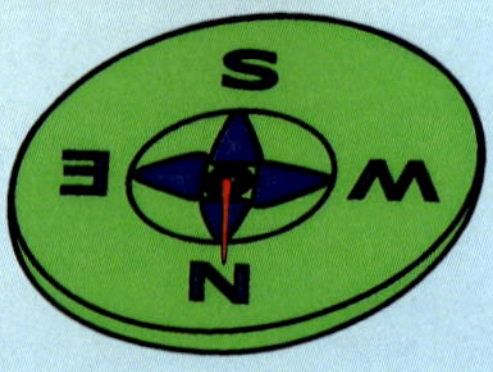

The **WEATHER**

The **Sideways Man** explained that really cold ***water*** freezes into ***ice*** (0°C) or lower. When water freezes into ice it expands as it becomes solid. Ice is lighter (less dense) than water and floats - just like icebergs on the earth's freezing polar caps.

Try this out:

- Half fill two identical cups with water
- Place one cup in the freezer overnight and the other one in the fridge
- In the morning compare the two cups.

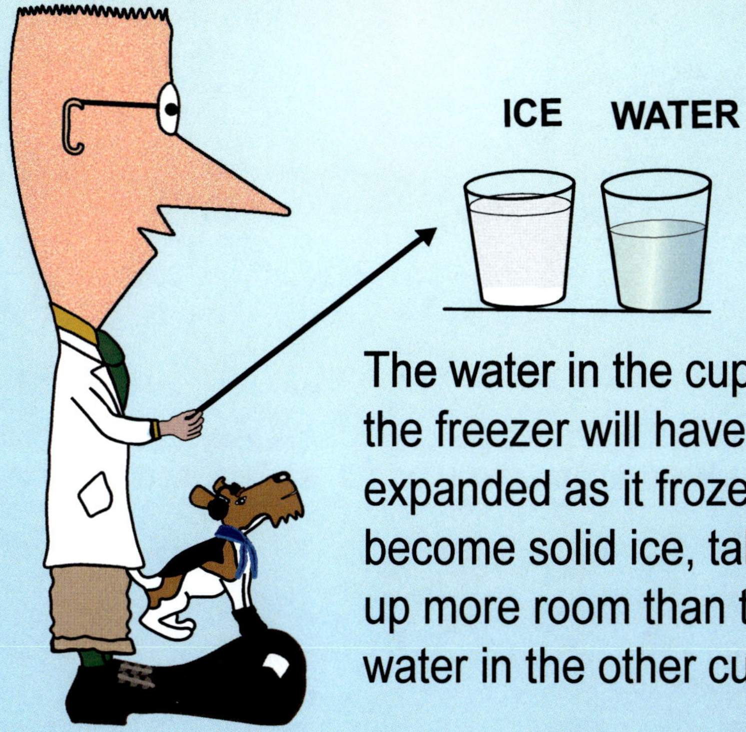

The water in the cup from the freezer will have expanded as it froze to become solid ice, taking up more room than the water in the other cup.

The **WEATHER**

The **Sideways Man** said we use ***thermometers*** to measure the ***temperature*** of our ***weather***. Temperature is measured in degrees celsius. A thermometer has liquid inside it which expands when the weather warms up on hot days and contracts when the weather gets colder.

Place a thermometer in hot water and measure the temperature and then place it in cold water. Watch as the liquid contracts to measure the temperature of the cold water.

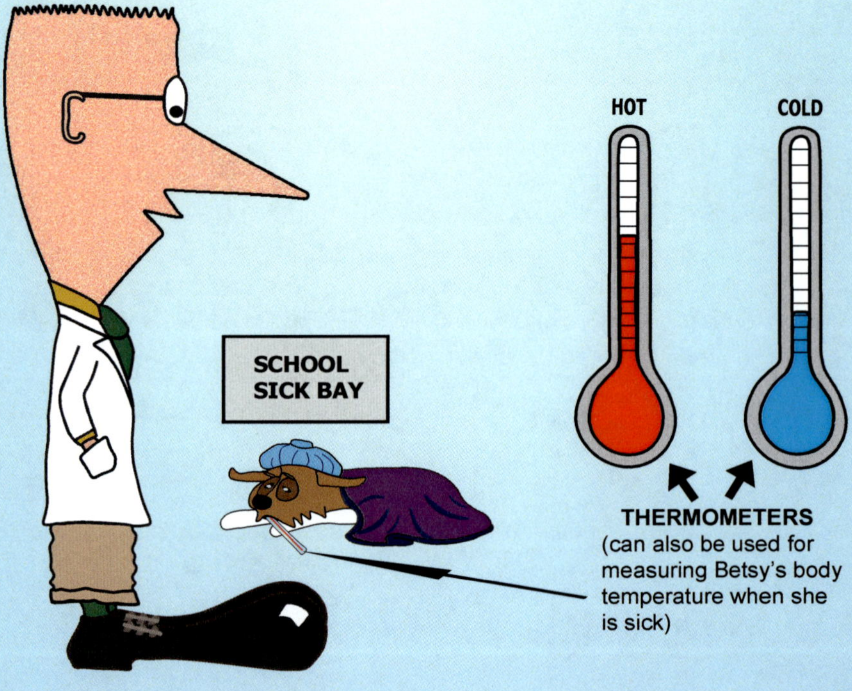

GRAVITY

Rockets are pushed along by blasting out a jet of gas made by burning rocket fuel. As the gas pushes out the base of the rocket there is an equal force pushing it to move the other way, against gravity and up into space.

Make your own balloon rocket: Thread some string through a straw. Tape the string across the room from down low to a high point. Then blow up a balloon and while holding it closed, tape it to the straw and let the balloon go and watch it blast along the length of the string.

GRAVITY

On Thursday the **Sideways Man** told us that we are surrounded by forces of gravity everywhere we go and in almost everything that we do. ***Gravity*** is a pulling force that keeps us on the earth.

When we drop something, it always falls downward because of gravity.

Try it with a slinky.

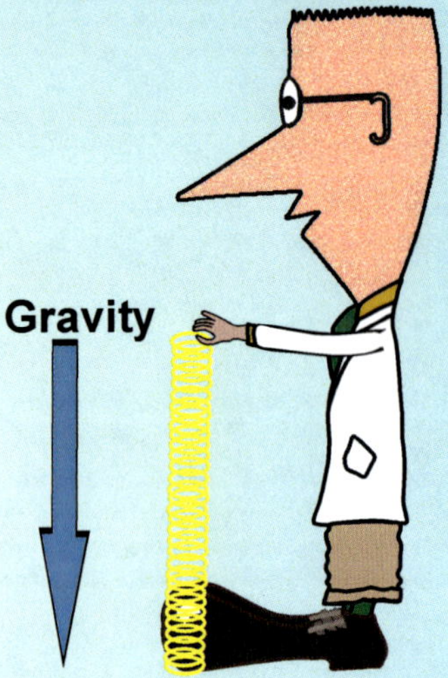

Also, every object has a centre of gravity but if we get out of line with our centre of gravity, we can lose balance.

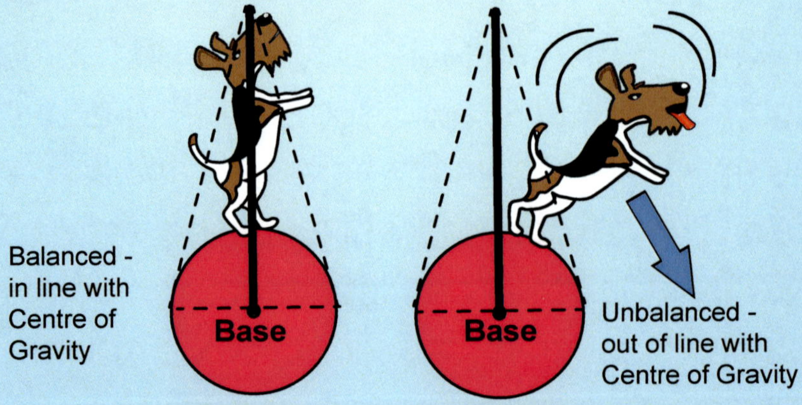

GRAVITY and FLIGHT

Airplane wings are shaped to make air move faster over the top of the wing. When air moves faster, the pressure of the air decreases. So the pressure on the top of the wing is less than the pressure on the bottom of the wing. The difference in pressure creates a force on the wing that ***lifts*** the wing against gravity and the airplane flies up into the air.

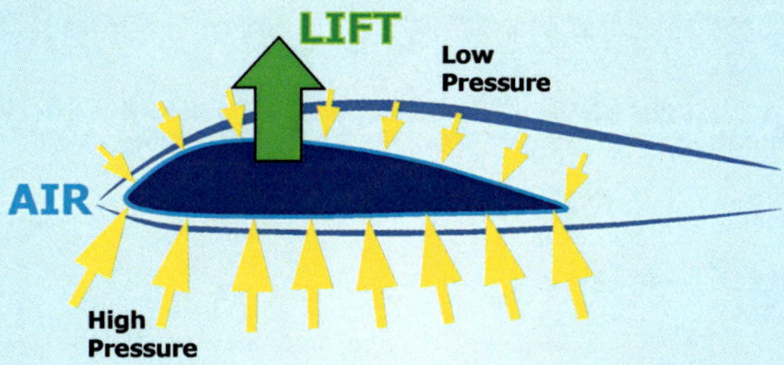

That is how birds fly, too.

FLOATING

On Friday the **Sideways Man** showed us that a boat ***floats*** if it is ***lighter*** than the same volume of water it displaces. The water beneath the boat pushes up (buoyancy) to balance the weight of the boat pushing down (gravity).

Big ships float because they displace so much water and the amount of air inside them is ***lighter*** than the same volume of water being displaced. On its own a rock will sink because it is ***heavier*** than the water it displaces.

QUESTIONS

The **Sideways Man** always says to make sure you ask questions about what you see all around you, like:

1 What makes the sky blue?
2 What is electricity?
3 What makes the tides change in the sea?
4 What is a robot?
5 What makes stars shine bright at night?
6 How do fish breathe under water?
7 What is a volcano?
8 What is combustion?
9 What is acceleration?
10 Why does Earth have different temperatures?

Then, ask your teachers or a parent to explain your questions. Look up the answers in the library, or on the internet, to learn more about them.

BUT there is one question that will always remain unanswered:

WHY IS IT THAT I ONLY EVER REMEMBER SEEING THE SIDEWAYS MAN - SIDEWAYS?

Write up your favourite Experiment Laboratory Report here:

1. Experiment Goal

2. Experiment Method

3. Experiment Results

4. Laboratory Report written by:

Your Name   Date

Can you draw the **Sideways Man**?

TSM
BYE BYE!

Made in the USA
Las Vegas, NV
20 March 2026

43946199R00019